Fuse works

Understand the structure functions and composition of fuse dissection

Dr Joe smith

Contents

Chapter1

Introduction to dissection of fuse

Dissection of fuse is a process of carefully separating the different parts of a fuse in order to understand its structure, function and composition. Fuses are important components in electrical circuits as they provide protection against excessive electrical currents. As the saying goes, "knowledge is power", understanding the anatomy of a fuse can help individuals to make informed decisions while selecting the appropriate fuse for a particular circuit. The word 'dissection' often evokes images of biology classes, where students cut open animals to study their anatomy. Similarly, the process of dissection of fuse involves cutting open the fuse, separating its various components and examining them closely. This process can be carried out on different types of fuses, like ceramic fuses, glass fuses, cartridge fuses, etc. Each type of fuse has a distinct design and composition, and the dissection process allows individuals to understand the differences and similarities between them. The dissection of fuse can be a fascinating and informative experience for anyone interested in electronics and electrical circuits. It offers a hands-on approach to learning and can enhance one's knowledge and understanding of the workings of a fuse. It can also be a valuable tool for professionals like electricians and engineers who work with fuses regularly and need to have a thorough understanding of their structure and function. The main component of a fuse is the fuse element, which is made of a conductive material such as copper or silver. This element is the key player in the protection mechanism of a fuse. The dissection process allows one to examine this element closely and understand its properties that enable it to handle high electrical currents. The size and thickness of the fuse element can vary, depending on the type and rating of the fuse, and this can affect its performance under different conditions. Another important aspect of a fuse is its housing. This is the outer covering of the fuse that protects it from external factors like moisture, dust, and physical damage. In the dissection process, one can observe the materials used for making the housing and explore their durability and suitability for different environments. For example, ceramic fuses have a ceramic housing that provides better protection and insulation than glass fuses with their more fragile glass housing. Apart from the fuse element and housing, the dissection process also sheds light on other components like the metal end caps, fuse link, and filler material. These elements play significant roles in the functioning of a fuse, and understanding their properties can help in selecting the right fuse for a specific circuit. For instance, the filler material used in the fuse can affect its response time to overload conditions. A filler material with higher resistance will take more time to fuse compared to one with a lower resistance. One of the essential aspects of the dissection of fuse is safety precautions. As fuses are designed to handle high voltages and currents, it is crucial to follow proper safety measures while handling them. This includes wearing appropriate protective gear like gloves and safety glasses, working in a well-ventilated area, and using specialized tools for the dissection process. These precautions ensure the safety of the individual performing the dissection and prevent any damage to the fuse during the process. Dissection of fuse also helps in identifying any defects or damages in the fuse. By examining its various components, one can determine if the fuse has been subjected to high currents or if there is any physical damage that may affect its performance. This information can be useful for troubleshooting electrical problems and identifying the cause for fuse failure. It can also assist in determining the appropriate replacement fuse for the circuit.

Chapter2

Parts of fuse

Fuse Body – This is the outer casing of the fuse, which contains all the internal components. The body is usually made of ceramic, glass, or plastic, depending on the type of fuse. Ceramic fuses are better suited for high voltage applications, while glass and plastic fuses are more suitable for low voltage applications. 2. Fuse Element – This is the most important part of the fuse, responsible for regulating the current flow. It is a thin strip or wire made of materials such as alloys, copper, or silver. When the current exceeds the rated capacity, the element heats up and melts, breaking the circuit and preventing any further damage. 3. End Caps – These are metal caps on each end of the fuse body, which provide the connection between the fuse and the circuit. The end caps are usually made of nickel, brass, or copper, which are good conductors of electricity. They also play a role in holding the fuse element in place. 4. Arcing Material - As the fuse element melts due to excessive current, it produces arcs of electricity. This arcing material, usually made of sand or quartz, is placed inside the fuse body to absorb and quench these arcs. It helps in preventing the fuse from exploding and causing further damage. 5. Filler Material – Fuses are filled with a non-conductive filler material, such as sand or quartz, to prevent the air from entering and igniting the arcs produced by the melting fuse element. The filler material also helps in dissipating the heat generated during the melting process. 6. Voltage Rating – The voltage rating indicated on the fuse specifies the maximum voltage that the fuse can handle. It is crucial to use the correct voltage-rated fuse for a particular circuit, as using a lower-rated fuse can result in incomplete protection, while using a higher-rated fuse can lead to damage or hazards. 7. Current Rating – The current rating of a fuse determines the minimum and maximum levels of current that the fuse can carry safely. It is important to choose a fuse with the appropriate current rating for a specific circuit to ensure effective protection.

Using a fuse with a lower current rating can result in frequent blows, while a higher-rated fuse may not blow even when the current exceeds the safe limit. 8. Blowing Time - The blowing time of a fuse refers to how quickly it breaks the circuit after the current exceeds its rated capacity. It is an essential factor to consider while selecting a fuse for a specific application. Fast-acting fuses have a quick blowing time, suitable for devices with sensitive electronic components, while slow-blow fuses take more time to blow, suitable for applications with inductive loads. 9. Indicator - Some fuses come with a built-in indicator, such as a filament or a colored strip, to help identify when the fuse has blown. This makes it easier to troubleshoot and replace the fuse, saving time and effort. 10. Fuse Holder – A fuse holder is a device used to mount and protect the fuse. It provides a secure connection between the fuse and the circuit, preventing any accidental touch or tampering. There are different types of fuse holders available, such as cartridge, panel mount, and plug-in holders, depending on the type of fuse and the application. 11. Resettable Fuse – Unlike traditional fuses, which need to be replaced once blown, resettable fuses have a mechanism to restore the circuit after blowing. These fuses have a built-in circuit breaker, which cuts off the circuit when there is an overload and gets back to its normal state after the fault is removed. 12. Surge Protector –

Some fuses come with a built-in surge protector to protect the device and the circuit from voltage spikes or surges. This is particularly useful in areas with frequent power fluctuations.

Chapter 3

Fuse and how it operates

A fuse is a safety device that is designed to break the circuit in case of a sudden surge of current, which can cause damage to the circuit or the device. It is a one-time use protective device that must be replaced after it has blown. In simple terms, a fuse is like a safety valve that detects an abnormal flow of electricity and prevents any further damage. A fuse is made up of a thin strip of metal, typically copper or aluminum, enclosed in a heat-resistant material such as glass or ceramic. This strip of metal is called the fusible element or the fuse link and is the most critical component of the fuse. It is designed to melt and break the circuit when exposed to excessive heat, preventing any further damage. Initially, fuses were made using lead or other metals that had low melting points, making them quick to respond to high current flow. However, with technological advancements, fuses are now made using conductive alloys that are tailored to melt at specific temperatures to suit different electrical applications. The two ends of the fusible element are connected to metal caps, which are in turn connected to the electrical circuit. These caps are usually made of a conductive material like brass, which ensures good electrical connection. The caps are then encased in a non-conductive housing, usually made from glass or ceramic, to protect the fuse from any external damage. Fuses come in a variety of sizes and shapes, depending on the application and the amount of current they are designed to handle. The most commonly used type of fuses is the cartridge fuse, which is cylindrical in shape and is used in high-voltage applications. Another popular type is the blade fuse, which is used in low-voltage applications, like those found in automobiles. Now that we understand the components of a fuse, let's take a closer look at how it operates. When the electrical current exceeds the safe limit, the temperature of the fusible element starts to rise. As the temperature rises, the metal strip begins to melt, and when it reaches its melting point, it breaks the circuit and stops the flow of electricity. The melting point of the fusible element is carefully calibrated depending on the amount of current that the fuse is designed to handle. If the current exceeds the limit for an extended period, the fuse will eventually melt and break the circuit. This process happens within a fraction of a second, preventing any potential hazards. Once a fuse has blown, it must be replaced to restore the circuit's functionality. This is why fuses are commonly referred to as one-time-use devices. However, some modern fuses come with a resettable feature, where the fuse element can be repaired or reset after it has blown. These resettable fuses are commonly used in electronic devices like computers, where replacing the fuse is not a practical option. Fuses play a significant role in preventing electrical fires and other hazards, but they also have some drawbacks. As mentioned earlier, fuses are one-time-use devices, which means they need to be replaced every time they blow. This can be inconvenient and can also increase maintenance costs, especially in industrial settings. Moreover, traditional fuses do not provide any indication of when they are about to blow. This issue has been addressed with the introduction of thermal fuses, which come with a built-in mechanism that changes color when the temperature rises, giving a visual indication of the fuse's status. Another limitation of fuses is that they do not differentiate between normal and abnormal current spikes. This means that in case of a temporary surge in current, the fuse will blow, disrupting the circuit unnecessarily. To address this, circuit breakers are often used in place of fuses. Circuit breakers can be reset after they trip, and they respond more accurately to changes in current flow.

chapter4

how does expulsion fuse works

An expulsion fuse is an electrical safety device that is used to protect a circuit from excessive current. It works by breaking the circuit and interrupting the flow of electricity, thus preventing any potential damage due to overheating or short circuits. The basic principle behind the functioning of an expulsion fuse is the transfer of energy. When excessive current flows through the fuse, the heat produced causes a thin layer of silver alloy to melt. This layer is coated on a copper or aluminum strip, which acts as a conductor. The melting of the silver alloy creates an arc between the two ends of the strip, which leads to the interruption of the circuit. The design of an expulsion fuse is quite simple yet effective. It consists of a wire or strip of metal that is enclosed in a metal or ceramic tube. This tube is filled with a filler material, usually made of quartz sand. The filler material serves two essential purposes – to provide physical support and to quench the arc formed during fuse operation. The expulsion fuse is connected in series with the electrical circuit and is designed to handle a specific level of current. If the current surpasses the rated value of the fuse, it will melt and break the circuit, thereby protecting the devices and components connected to it. One of the major advantages of an expulsion fuse is its ability to handle high currents for a short period. This makes it an ideal protection device for equipment such as motors, transformers, and generators, which may experience high starting currents. Unlike other safety devices such as circuit breakers, which can take time to sense and interrupt the current, an expulsion fuse can act swiftly and prevent damage to the equipment. Another advantage of an expulsion fuse is its simplicity and easy installation. It does not require any external control or power source to function, which makes it a reliable and cost-effective safeguarding device. It also does not involve any moving parts, making it less susceptible to wear and tear. The operation of an expulsion fuse can be explained in three stages – melting, vaporization, and extinction. The first stage occurs when the level of current exceeds the rated value of the fuse. The heat produced due to the high current causes the silver alloy coating to melt, which leads to the formation of an arc. In the second stage, the heat from the arc causes the filler material to vaporize, depleting the oxygen in the tube, and hence extinguishing the arc. Finally, in the third stage, the gas pressure rises, and the arc is quenched, breaking the circuit. Apart from the basic expulsion fuse, there are two other types – the bottle type and the helix type. The bottle type has a larger volume of filler material, which makes it more suitable for high voltage and high-current applications. The helix type is designed for high voltage and low current applications and has a compact design that allows for easy mounting. Expulsion fuses are also designed with different time-current characteristics to suit different applications. The time-current characteristic of a fuse determines how long it can operate under excessive current conditions before interruption. This feature is essential, especially for equipment with varying inrush currents, as it ensures that the fuse does not blow off unnecessarily. Despite its advantages, expulsion fuses have some limitations and drawbacks. One of the major drawbacks is that they cannot be re-used once they have blown off. They need to be replaced every time the circuit they are protecting experiences overcurrent. This can result in downtime and additional costs, especially in industrial applications.

Chapter5

What is inside a fuse

The first and most important component of a fuse is the element, which is the thin wire mentioned earlier. This wire is usually made of a metal with a low melting point, such as copper or silver. The melting point of the wire is chosen carefully, based on the amount of current that the circuit can safely handle, so that it will melt when the current exceeds the safe limit. The element is responsible for breaking the circuit in case of an overload, preventing the excess current from flowing through and damaging the circuit. The element is enclosed in a non-conductive material, typically glass or ceramic, for protection and insulation. The glass or ceramic tube is usually filled with an inert gas, such as nitrogen, to prevent the element from burning or exploding when it melts. This also helps to contain any potential fire hazard within the fuse. On either end of the element, there are two metal caps called ferrules, which are used to connect the fuse to the electrical circuit. These ferrules are usually made of copper and are designed to be easily connected to other electrical components such as wires or terminals. The metal caps are designed to conduct electricity and also provide a secure connection to the circuit. The main purpose of a fuse is to safeguard electrical circuits from overloads, short circuits or faults. When there is a sudden surge or spike in the electrical current, the fuse's element will heat up and eventually melt, thanks to its low melting point. This disconnects the circuit and prevents any further flow of current. This is why fuses are often referred to as "one-time use" devices. Once the element melts and breaks the circuit, the fuse needs to be replaced with a new one. There are different types of fuses, each designed for a specific purpose. The most common types include glass cartridge fuses, plug fuses and blade fuses. The glass cartridge fuse is the traditional type with a glass tube and visible wire element, often used in older buildings. Plug fuses can be found in the wiring of many homes and have a similar appearance to traditional glass fuses. They are designed to be screwed into a fuse socket, and their element is not visible. Blade fuses are the newer type of fuse that are typically used in modern vehicles and electronic equipment. They have a plastic body and two metal blades that fit into the fuse slot. The rating of a fuse is an important factor to consider when using one. It refers to the maximum amount of current that a fuse can safely handle without melting. The ratings are usually printed on the fuse or indicated by colors on the end caps, making it easy to identify the correct rating for a particular circuit. Choosing the wrong rating can cause the fuse to either blow too quickly, even with normal current, or not blow during an overload, effectively defeating its purpose. In addition to the basic components, some fuses may also have other features such as a time delay, known as a time-lag fuse or a slow-blow fuse. These types of fuses are designed to handle brief increases in current without blowing, making them suitable for devices that have a high starting current, such as motors or refrigerators. Time-delay fuses can also handle surges during normal operation, preventing unnecessary tripping. However, in cases of severe overloads, they will blow just like any other type of fuse.

Dissection of fuse diagram

A fuse diagram is a visual representation of the fuses and relays located in a specific electrical system in a vehicle. It provides a detailed breakdown of each fuse and its function, along with the location and amperage rating. This type of diagram is essential for troubleshooting electrical issues and identifying faulty fuses. There are several different types of fuse diagrams, including those for the main fuse box located in the engine compartment, and those found in the interior fuse panel. In this dissection, we will take a closer look at how to read and understand a fuse diagram, and the importance of having one in your vehicle. Reading a fuse diagram is relatively straightforward, but it may seem overwhelming at first. Most diagrams are color-coded and have symbols and numbers, making it easier to identify the different components. The first step is to locate the legend or key, which will explain the color and symbols used in the diagram. Once you have identified the legend, you can start to examine the diagram. Fuses are typically labeled with a number, corresponding to a specific electrical component, such as the radio, headlights, or wipers. The fuse rating, or amperage, is also indicated, which is important to note when replacing a blown fuse. The color of the fuse also signifies its amperage rating. For example, larger fuses with higher amperage may be colored yellow or red, while smaller fuses with lower amperage may be colored blue or green. This color-coding helps to prevent the wrong fuse from being installed, which could cause further damage to the electrical system. The next step is to identify the location of the fuse in the diagram. This is typically represented by a number or letter, which corresponds to the location of the fuse in the physical fuse box in the vehicle. This is crucial information, as it allows you to locate and replace a blown fuse in the correct position. However, not all fuses are labeled in the same way. Some diagrams may use abbreviations or acronyms, which can be more challenging to decipher. In this case, it is best to consult the vehicle's manual or an online resource for a clearer understanding of the diagram. In addition to fuses, a fuse diagram may also include relays. A relay is an electrical switch that helps to control a larger electrical component, such as the fuel pump or air conditioning system. Similar to fuses, relays also have their own ratings, which are indicated on the diagram. A fuse diagram can also provide valuable information about the vehicle's wiring and electrical layout. It may show how different components are connected and how they interact with each other. This can help diagnose and troubleshoot more complex electrical issues. Having a fuse diagram can save a lot of time and frustration when dealing with electrical problems in a vehicle. It allows for a quick and accurate identification of faulty fuses, which can be easily replaced to resolve the issue. It also prevents the need for trial and error, which can lead to further damage to the vehicle's electrical system. Another advantage of a fuse diagram is that it helps to prevent incorrect fuse replacement. If a fuse has blown, it is important to replace it with one of the same amperage rating. A higher or lower-rated fuse could either not provide enough power or cause an overload, leading to further damage.

The end